HMH

PHOTOGRAPHY CREDITS: **Cover** © TK; **1** © Westend61/Getty Images; **2** © Diane Randell/Alamy; **3** © blickwinkel/Alamy; **4** © Westend61 GmbH/Alamy; **5** © otsphoto/Shutterstock; **6** © by_nicholas/Getty Images; **7** ©Stephano Rossi/Alamy

Copyright © by Houghton Mifflin Harcourt Publishing Company

All rights reserved. No part of this work may be reproduced or transmitted in any form or by any means, electronic or mechanical, including photocopying or recording, or by any information storage and retrieval system, without the prior written permission of the copyright owner unless such copying is expressly permitted by federal copyright law. Requests for permission to make copies of any part of the work should be submitted through our Permissions website at https://customercare.hmhco.com/contactus/Permissions.html or mailed to Houghton Mifflin Harcourt Publishing Company, Attn: Intellectual Property Licensing, 9400 Southpark Center Loop, Orlando, Florida 32819-8647.

Printed in the U.S.A.

ISBN 978-1-328-77227-5

4 5 6 7 8 9 10 2562 25 24 23 22 21

4500844736   A B C D E F G

If you have received these materials as examination copies free of charge, Houghton Mifflin Harcourt Publishing Company retains title to the materials and they may not be resold. Resale of examination copies is strictly prohibited.

Possession of this publication in print format does not entitle users to convert this publication, or any portion of it, into electronic format.

Count the puppies 1 to 10.
Now hide three puppies
and count again.

2   How many puppies are left?

Eight hungry puppies love to eat.
Hide two puppies
from noses to feet.

Count how many puppies are left. 3

Seven puppies play in the hay.
Hide three little puppies
who are sleepy today.

How many are left?

Five puppies walk at the shore.
Hide three or four
or even more.

How many did you hide? How many are left?

Two little puppies nap in the sun.
Hide them both.
They had too much fun.

How many are left after you subtract 2?

One little puppy sits all alone.
There are no puppies to play with,
so he runs for home.

How many puppies are left?

Math Concepts

Hungry Puppies

Draw

Look at page 3. On a piece of paper, draw a W for each puppy that is white. Draw a B for each puppy that is brown.

Tell About

Monitor/Clarify Look at page 3. Tell how many puppies are white. Tell how many puppies are brown. Tell how many puppies there are in all.

Write

Look at page 3. Write how many puppies there are in all.